nderstanding Electricity for A Level

torial for the electricity component of OCR B and AQA AS level P

re are of course dozens of text books, revision guides and study a........ .... teaching this subject over 30
rs, however, I believe that I communicate my understanding very well and that I am familiar with many of the
conceptions that students have which can hinder their understanding. I very much hope you enjoy this book and
it useful. Please leave a comment to let me know what you think. I plan to produce similar tutorials in the future
uding Understanding Waves and Understanding Mechanics so keep an eye out for them.

David Drumm

1. Understanding Current   2
2. Understanding Voltage   8
3. Understanding Resistance   16
4. Electrical Power   19
5. Resistor Networks   22
6. More Components   27
7. Potential Dividers   31
8. Internal Resistance   35
9. Resistivity   39
   Questions   43   Answers   49

ore we start

h help you to tackle exams on electricity, if you're willing to do the work, but for a real understanding of what's
1g on you need to do some practical work. By building circuits and actually handling components you will
rnalise the concepts covered here. One of the most useful things is figuring why things don't work. I have learnt a
at deal about electricity from pondering over why a particular bulb doesn't light up or why a voltmeter doesn't
the reading that it should. There are a number of kits available from educational websites, e.g. philipharris.co.uk
betterequiped.co.uk. If you have not actually done any of the practical work necessary to support your course
will be at a big disadvantage and I would recommend investing in a modest amount of equipment.

teaching advanced level electricity here, i.e. for A level exams. I will not assume a great deal of prior knowledge
ever I will assume some mathematical skill, in particular that you can use a calculator. I will also assume that you
w that 0.43MJ is the same as 430kJ and the same as 4.3×10^5J.

re will be a lot of quantities with symbols and units you need to learn as well as equations expressing the
tionships between them. I will assume that you have the mathematical skill to rearrange these equations as
ded. You **MUST** make an effort to learn these equations. You do not have time in an exam to hunt through a
nula sheet for an equation that probably isn't in there anyway.

ach this topic to students over a period of about 3 or 4 weeks so my advice would be to take your time going
ugh this material. Certainly do not go onto a section until you fully understand the one before.

ck your exam board's specification to make sure you are not learning stuff you don't need, e.g. AQA Physics does
include conductivity but OCR does.

1. Understanding Current

An Analogy for Electrical Current

Imagine a flowing river.

When we talk about the current in a river what do we mean?

What would the current in a river be measured in?

The current in a river is the **amount of water** which flows past any point in a certain time. In S.I. units this would either be measured in kilograms per second or metres cubed per second (mass flow rate or volume flow rate).

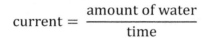

$$current = \frac{amount\ of\ water}{time}$$

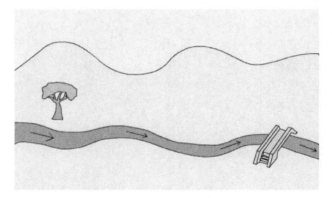

Note the following:

Current is **not** the speed of the water. The speed of the water in a narrow fast flowing river may be faster than in a much wider river however the current may be greater in the wider deeper river.

If 150kg of water flow past the tree every second how much will flow under the bridge every second?

Of course it will be the same. We are not gaining or losing any water (unless it is raining or there is a cow having a drink between them).

Why does a current flow?

A current will only flow downhill so as it flows the water loses gravitational potential energy.

Now let's consider a simple electrical circuit

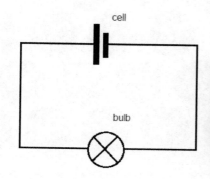

What flows through the wires in an electrical circuit?

An electrical current is the flow of **charge**

$$\text{current} = \frac{\text{amount of charge}}{\text{time}}$$

What is charge?

This is a very hard question to answer. It is a property that some things have and there are two different kinds of it which we call positive and negative. An object is either positively charged, negatively charged or is neutral.

We give charge the symbol Q and the amount of charge is measured in Coulombs which has the symbol C.

If current, which has the symbol I, is the flow of charge then current could be measured in Coulombs per second. It has its own units however Amps or Amperes which has the symbol A.

So far we have ...

Quantity	Symbol	Units	Symbol for units
Charge	Q	Coulombs	C
Time	t	Seconds	s
Current	I	Amps	A

So charge flows through the wires and components in electrical circuits.

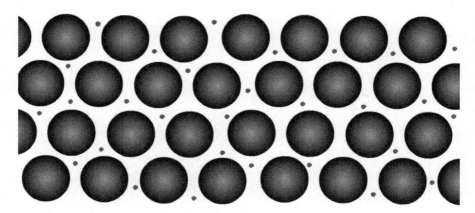

This diagram represents the atoms in a copper wire. In between the atoms are tiny particles called free electrons or delocalised electrons. In a metal there are LOTS of them.

Each electron has a tiny negative electrical charge of -1.6×10^{-19}C. We give this tiny amount of charge the symbol "e" and it will come up again and again as you study physics.

The electrons are not just sat there. They are bouncing around like crazy between the atoms which are also jiggling around quite a bit (depending on the temperature of the metal).

When a current flows through the wire (why it would flow we'll worry about later) these electrons drift in a particular direction. They would like to accelerate and move very fast but keep colliding into the jiggling copper atoms, like balls falling downstairs. So they bounce and bounce and squeeze through the gaps actually travelling at an average velocity of only about a few mm per second.

A couple of interesting points:

Why is it harder for a current to flow when the metal is hotter?

Why do metals get hotter when a current flows through them?

Well if the metal is hotter the atoms making it up are jiggling around more. This makes it harder for the electrons to get through the gaps between them as there are more collisions. Imagine you go to a party and have to try to cross the dance floor. It's a lot easier during a slow song than if everybody is dancing like crazy.

And what effect does it have on the atoms when these electrons keep bashing into them? Well they are going to gain energy and vibrate more. When a current flows through a metal the temperature of the metal increases.

Note that current is NOT the flow of electrons. It is the flow of charge. That charge is usually carried by electrons but sometimes it isn't, e.g. in electrolysis the charge is carried by positive and negative ions in solution.

The direction of the current is actually defined as the direction of the flow of positive charge, opposite to the direction that electrons flow in. Don't worry too much about this at this stage but it will crop up later.

ng the Current Equation

s get back to our equation which we can now write using the correct symbols

$$\text{current} = \frac{\text{amount of charge}}{\text{time}} \qquad\qquad I = \frac{Q}{t}$$

Q. A current of 0.7A flows through a component for 5 minutes.

a) How much charge flows through the component in this time?

b) How many electrons carry this much charge?

rranging the equation gives $Q = I\,t$ = 0.7 x (5 x 60)   = 210C

ectron has a charge of -1.6×10^{-19}C so 1 Coulomb would be carried by the reciprocal of this,

$\frac{1}{1.6 \times 10^{-19}}$ electrons.   210C would be carried by this x 210, i.e. 1.31×10^{21} electrons

e that this is a HUGE number. Despite the fact that the electrons are drifting quite slowly there are a HUGE
nber of them in the wire, about one for every metal atom. Metals are good conductors because the number of
lable charge carriers per unit volume is so high.

es and Parallel Circuits

rent is measured using an ammeter. To measure the current through a component we put the ammeter in series
it. This simply means before or after it. Note that the current is therefore the same going into and out of a
ponent. Current does not get used up.

current in a series circuit is the same at any point. Remember the tree and the bridge?

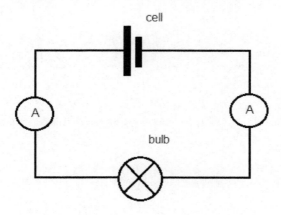

In a parallel circuit the current splits but the total current must stay the same. Imagine if two rivers joined together
The current would be the sum of the currents in both the tributaries.

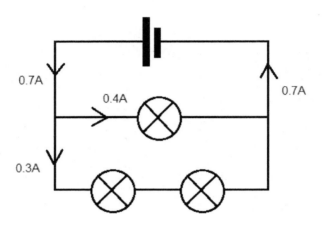

Kirchhoff's first law tells us:

The total current entering any point must equal

the total current leaving that point

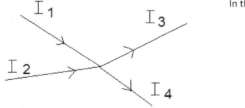

In this diagram

$$I_1 + I_2 = I_3 + I_4$$

$$\sum I_{in} = \sum I_{out}$$

This is a very simple statement but can be very powerful for solving certain types of problem.

mmary of Part 1

rrent is the rate flow of charge

rrent, charge and time are related by the equation $I = \dfrac{Q}{t}$

arge in metal wires is carried by electrons

e electron has a charge of $-1.6 \times 10^{-19}C$

measure the current through a component we put an ammeter in series with it

e current at any point in a series circuit is the same at any point

a parallel circuit the current splits but the total current must stay the same

2. Understanding Voltage

What makes a current in a circuit flow?

The power supply does. Without the power supply there would be no current.

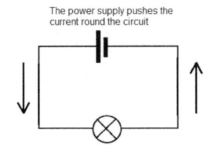

The power supply pushes the
current round the circuit

If we increased the voltage of the power supply then the current would be bigger. It is useful to think of the voltage of a power supply as how hard it pushes the current. To measure the voltage across a power supply we would use voltmeter. Remember that voltmeters always go in parallel.

If we put cells in series we create a battery and, if we connect them properly, we get a bigger voltage.

1.5V   3V   4.5V

Several cells pushing together results in a bigger push.

Note: Current goes **through** things. Voltage goes **across** things. If we put a voltage **across** something then a curren flows **through** it.

The circuit below would be used to investigate how the current through a component, such as a lamp, changes wh we change the voltage across it. The component at the top is a variable d.c. power supply. There are other ways o changing the voltage across the component which we consider later.

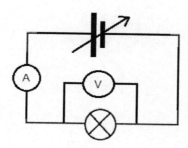

Imagine we changed the voltage from 0 to 5V and measured the current through a 6V bulb. We would get a graph shaped like this. (The actual values for the current would depend on the bulb)

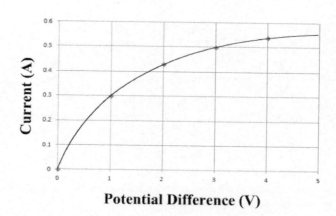

Notice that for a bulb the graph is not a straight line, i.e. the current is not proportional to the voltage.

What would happen if we used a negative voltage by connecting the power supply the other way round?

Well the current would flow in the opposite direction. Technically if we use a negative voltage then we'll get a negative current. The graph below is called the characteristic for a bulb. There are a few of these graphs that we will need to be familiar with.

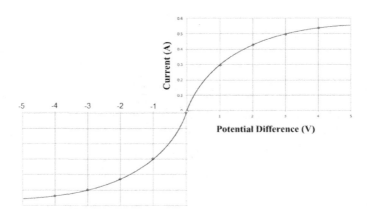

The current in a river flows downhill. As it flows the water loses gravitational potential energy.

This is very similar to what happens in an electrical circuit.

When a current flows through a component it loses electrical potential energy. This energy is transferred into other forms, e.g. heat.

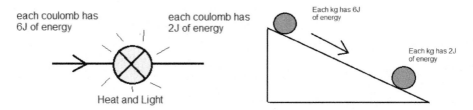

The amount of electrical potential energy each coulomb of charge has at any point in the circuit is called the **potential** at that point. Potential is measured in joules per coulomb.

The potential before the bulb in the diagram above will therefore be higher than the potential after the bulb. The difference in potential before and after the bulb is called the potential difference, or p.d., across it.

The potential difference across a component is also known as the voltage across the component

The voltage across a component is therefore the difference in electrical potential energy that each coulomb of charge transfers flowing through it.

$$\text{Potential difference} = \frac{\text{Energy transferred}}{\text{Charge}} \qquad V = \frac{E}{Q}$$

So where does this energy come from?

comes from the power supply. In a cell a chemical reaction takes place which results in the charge going through it
ining electrical potential energy.

he voltage across a power supply is the E.P.E. that it gives to each coulomb that flows
rough it.

Quantity	Symbol	Units	Symbol for units
Voltage	V	Volts	V
Energy	E	Joules	J
Charge	Q	Coulombs	C

re's another useful analogy.

electrical circuit is like a racetrack. Lots of cars go round and round at a steady speed using up their petrol. When
ey run out at the end of the lap they go through the pit stop where they load up on petrol.

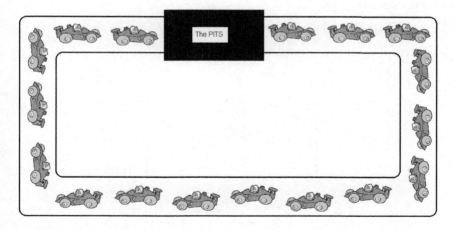

ch car represents a coulomb of charge

e amount of petrol given to each car represents the voltage of the power supply

e number of cars which pass a point every second represents the current in the circuit

What would the following represent?

a) A person by the side of the track measuring how many cars go past every second?

b) A person next to the pit stop measuring how much petrol is given to each car?

Using the Voltage Equation

A current of 0.7A flows through a bulb when a p.d. of 6V is put across it. How much energy will it transfer in minutes?

We know that the charge is given by $\quad Q = I\,t \quad$ = 0.7 x 5 x 60 = 210C

So the energy transferred $\quad E = Q\,V \quad$ = 210 x 6 =1,260J

One difference between questions at A level and those at GCSE is that there are often several steps required before you arrive at your answer. In this example we had to calculate the charge before we could work out the energy transferred. We weren't told to, we were expected to figure that out for ourselves.

The best advice I can give is to **LEARN YOUR EQUATIONS**. If you know all your equations like the back of your hand then it should be obvious what method you would use to get to what is asked for. You may be given a formula sheet in your exam but I have noticed that the students that rely heavily on these are the ones who have not made much of an effort before the exam and are doomed to failure.

Voltage in series and parallel circuits

Below is a circuit with two components in series, a bulb and a resistor.

What will the voltage across the resistor be?

6V

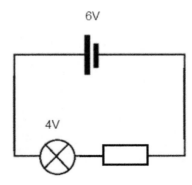

4V

It has to be 2V. If you understood the section about potential difference you should realise that each coulomb of charge is given 6J of energy by the power supply. If 4J of this is transferred by the bulb then this leaves 2J to be transferred as heat by the resistor.

sum of the potential drops in any closed loop must equal the power supply voltage

statement is actually just another form of the principle of conservation of energy.

vhat about parallel circuits?

What will the voltage across each component be in this circuit?

6V

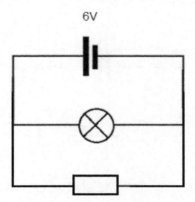

e's a useful tip. In this circuit we have two different branches. Whatever is happening in one branch does not
ct what is happening in the other. In other words, the fact that the resistor is there makes no difference to the
. The voltage across the bulb is therefore just 6V, as is the voltage across the resistor. This leads to another
en rule.

**nponents or combinations of components in parallel must have the same p.d. across
·m.**

What will the voltage across each component here be?

6V

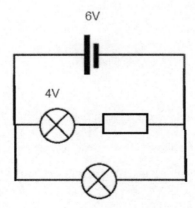

4V

resistor will have 2V across it as 4 + 2 = 6V

The second bulb will have 6V across it. The fact that there is another branch in the circuit makes no difference to the fact that it is in parallel with the power supply.

Here's one more example, one that students often find confusing.

What will be the voltage across each of these components?

6V

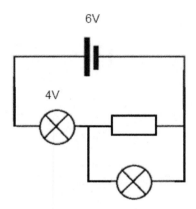

4V

If we follow our golden rules then the answer is pretty obvious.

The resistor and the second bulb are in parallel so they must have the same voltage across them. If we cover up the second bulb we can see that the resistor must have 2V across it as in that loop 4V + 2V = 6V. The answer is therefore that both the resistor and the second bulb have 2V across them.

Summary of Part 2

Voltage or Potential Difference is measured in volts

If we increase the voltage across a component then the current through it will increase

A graph of V against I is called the characteristic of a component

The V I characteristic of a bulb looks like this

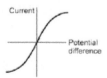

The voltage across a power supply is the amount of energy that it gives to each coulomb that flows through it

The voltage across a component is the amount of electrical energy transferred by each coulomb that flows though it

$$V = \frac{E}{Q}$$ Remember: 1 Volt = 1 Joule per Coulomb

In a series circuit the voltages across each component add up to the power supply voltage.

Components, or combinations of components, in parallel must have the same voltage across them

Kirchhoff's second law tells us:

The sum of the e.m.f.s in any closed loop must equal

the sum of the potential drops

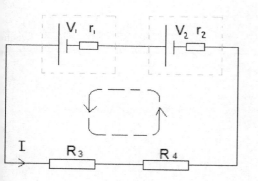

In this circuit

$$V_1 + V_2 = Ir_1 + Ir_2 + IR_3 + IR_4$$

$$\sum \varepsilon = \sum IR$$

In circuits with parallel branches there are a number of loops.

Applying the equation above, and Kirchhoff's first law, to these loops generates simultaneous equations which you can solve to find various unknowns.

3. Understanding Resistance

We have seen that the current through a bulb depends on the voltage we put across it. The more voltage we use the bigger the current.

What else does the current depend on?

There is a property of components called their resistance. Resistance R is measured in Ohms Ω

The resistance of a component is how hard it is for a current to flow through it.

Large resistance = Small current   **Small resistance = Large current**

You can also think of resistance as being the voltage you would need to put across a component to get a current of 1A to flow through it. Resistance is defined by this equation;

$$R = \frac{V}{I}$$

Quantity	Symbol	Units	Symbol for units
Voltage	V	Volts	V
Current	I	Amps	A
Resistance	R	Ohms	Ω

Remember or V I graph for a bulb.

Use the equation for above to calculate the resistance of the bulb when different voltages are put across it.

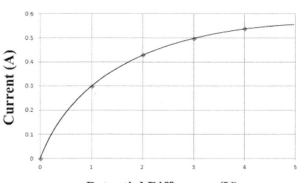

Potential Difference (V)

When the bulb has 1V across it its resistance = 1/0.3 = 3.33Ω

When the bulb has 4V across it its resistance = 4/0.54 = 7.41Ω

s resistance clearly increases as more current flows through it.

here are components which are designed to have a fixed resistance. These are called resistors.

he coloured bands on the resistor identify its value.

sing the Resistance Equation

ow much current will flow through a 47Ω resistor if we put 6V across it?

$$= \frac{V}{R} \qquad = 6 / 47 \qquad = 0.128A \text{ or } 128mA$$

What would a V I graph for a fixed resistor look like?

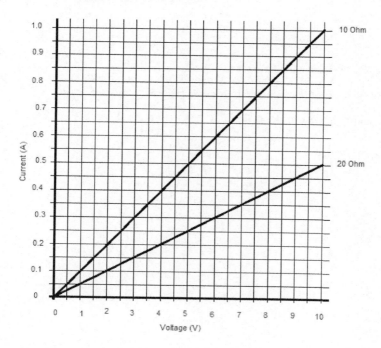

If the resistance is fixed then the value of V / I at any point must be the same. For a fixed resistor V is proportional to I. A fixed resistor therefore obeys Ohms law.

Note: **Do not learn that the resistance is equal to the gradient of the line**. This is only happens to be the case if the graph is a straight line. To find R we divide V by I.

Summary of section 3

Big resistance = Small Current   Small Resistance = Big Current

$$R = \frac{V}{I}$$

A graph of V against I for a component is called its V I characteristic

Ohm's law states that, at constant temperature, current is proportional to voltage

Fixed resistors obey Ohm's law. Bulbs do not.

4. Electrical Power

wer is the rate at which energy is transferred or the rate of doing work.

th men below can move that pile of manure from A to B but one of them can do it much quicker cause he is more powerful.

th of these kettles can boil 2 litres of water but the first kettle can do it quicker. It is more powerful.

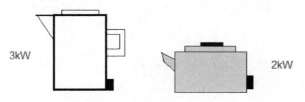

3kW

2kW

$$\text{Power} = \frac{\text{Energy Transferred}}{\text{time}} \qquad P = \frac{E}{t}$$

Quantity	Symbol	Units	Symbol for units
Power	P	Watts	W
Energy	J	Joules	J
Time	t	Seconds	s

ing the power equation

You need 4,200J to raise the temperature of 1kg of water by 1°C

a) *How much energy would be needed to boil 1kg of water starting at 20°C?*
b) *How long would it take to do this using a 3kW kettle?*

c) What assumption do you make in your calculation?

The energy needed would be 80 x 4,200 = 336,000J

The time would be $\quad t = \dfrac{E}{P} \quad = 336{,}000 \,/\, 3{,}000 \quad = 112s$

We are assuming that all of the electrical energy transferred is given to the water to heat it up, i.e. none lost to the surroundings.

Equations Equations Equations

We now have a number of equations at our disposal which will be useful tools for tackling any problem thrown at us. It is important that you learn them. You will not have time to look them up in an exam.

$$Q = I\,t \qquad E = Q\,V \qquad V = I\,R \qquad E = P\,t$$

By combining these equations we can generate a few more which are particularly useful for tackling cert kinds of problem.

$$P = V\,I$$

This is very useful for calculating the current drawn by a device from a power supply. When we know this we can quickly figure out what value fuse to put in its plug.

$$P = I^2 R$$

A very useful equation for calculating the rate of energy transferred into heat by a resistor. I often refer t power loss as I^2R.

$$E = V\,I\,t$$

Again a useful equation for calculating energy transferred

Summary of section 4

Power is the rate of transfer of energy, i.e. how much energy is transferred per second

$$P = \frac{E}{t} \qquad P = V\,I \qquad P = I^2 R$$

Power is measured in watts or kilowatts

5. Resistor Networks

We can talk about the resistance of a single component or we can talk about the resistance of a combination of components, e.g. a number of resistors in series or parallel.

Resistors in series

What would the resistance of this combination of resistors be?

The answer is simply 47Ω

For resistances in series the combined resistance is given by

$$R_{total} = R_1 + R_2 + ...$$

Resistors in Parallel

What would the resistance of this combination of resistors be?

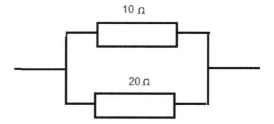

Before I give you the equation to work it out I would like you to consider this. The resistance must be less than 10Ω.

A 10Ω resistor would have a certain current flowing through it. By adding another resistor in parallel, no matter what value, we have provided another route for current to flow so the total resistance must be less

For resistances in parallel the total resistance is given by

$$\frac{1}{R_{total}} = \frac{1}{R_1} + \frac{1}{R_2} + ...$$

When using this equation you will find the button on your calculator labelled x^{-1} very useful

he example above $$\frac{1}{R_{total}} = \frac{1}{10} + \frac{1}{20}$$ which gives R as 6.67Ω

ou get a more complicated network such as the one below just simplify it by working out the resistance of parts
he circuit first.

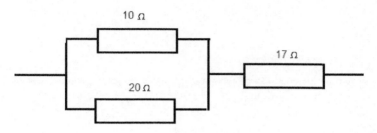

resistance of this network would be 6.67 + 17 = 23.7Ω

meters and Voltmeters

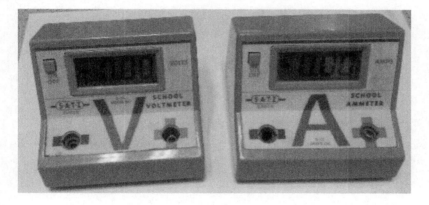

erfect ammeter has zero resistance. The resistance of ammeters is usually very small and so in nearly all
estions, certainly the ones in this book, you will assume that their resistance is zero.

Why should it be zero?

have seen that adding a resistance in series always increases the total resistance. If an ammeter did not have
o resistance then it too would add to the resistance of the circuit and therefore affect the current we want to
asure.

erfect voltmeter has infinite resistance. The resistance of voltmeters is usually very high (at least a MΩ) so in
rly all questions, certainly the ones in this book, we will assume that they have infinite resistance.

Why should it be infinite?

We have seen that adding another resistor in parallel always decreases the total resistance of the combination. This would therefore affect the voltage that we are trying to measure. The higher the resistance we add in parallel the less the total resistance is affected and so the less the voltage changes.

Conductance

Conductance is the opposite of resistance

Large conductance = Large current   **Small conductance = Small current**

Conductance G is measured in Siemens which has the symbol

Quantity	Symbol	Units	Symbol for units
Resistance	R	Ohms	Ω
Conductance	G	Siemens	S

To calculate the conductance of a component we would use

$$G = \frac{1}{R}$$ or we could use $$G = \frac{I}{V}$$

So why bother with conductance when we have resistance?

One reason we bother with it is because it's on the specification so we need to know it!

It does have some uses. If you think of a resistor as a conductor instead then the equation to work out the combined conductance of a combination is a bit easier. It avoids using the fiddly equation for resistors in parallel.

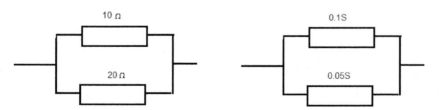

The combined conductance of several conductances in parallel is given by

$$G_{total} = G_1 + G_2 + \ldots$$

So in the example above it would be 0.1 + 0.05 = 0.15S

Summary of section 5

The resistance of a combination of resistances in series is given by

$$R_{total} = R_1 + R_2 + \ldots$$

The resistance of a combination of resistances in parallel is given by

$$\frac{1}{R_{total}} = \frac{1}{R_1} + \frac{1}{R_2} + \ldots$$

We assume that ammeters have zero resistance

We assume that voltmeters have infinite resistance

Conductance is the opposite of resistance

$$G = \frac{1}{R} \quad \text{or} \quad G = \frac{I}{V}$$

For components in parallel $\qquad G_{total} = G_1 + G_2 + \ldots$

6. More Components

There are several other components we need to be familiar with

The Diode   **The Thermistor**   **The L.D.R.**   **The Variable Resistor**

We will consider the characteristics of each of these in this section then in the next we will look at how they can be combined with other components to make useful circuits. We are not concerned with how they actually work at this stage.

The Diode

A diode only lets current flow one way.

- When it is "forward biased" it has a low resistance and a current will flow.
- When it is "reverse biased" it has a very high resistance and virtually no current flows.

Below is the V I characteristic for a diode. Note that when a negative voltage is put across the diode no current flows.

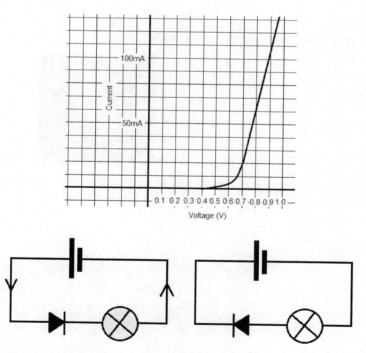

One specific use for diodes is in circuits called rectifiers. These change an a.c. current into a d.c. current.

Light emitting diodes, LEDs, are becoming more and more common as they are cheap and much more efficient than filament lamps.

The Thermistor

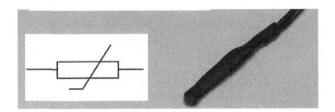

The resistance of a thermistor depends on its temperature

For a n.t.c. thermistor (negative temperature coefficient) then its resistance gets less when it gets hotter. Remember that this is the opposite of what happens to metals.

The L.D.R.

The resistance of an LDR depends on the intensity of the light falling on it

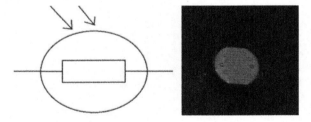

The more light falls on it the lower its resistance. LDRs are used in circuits where you want something to happen depending on how light or dark it is, e.g. turning on street lights automatically.

The Variable Resistor or Rheostat

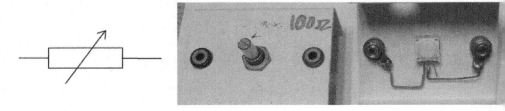

en you turn a knob its resistance gets bigger or smaller. One use is for a dimmer switch. Increasing
resistance in the circuit decreases the current so the bulb gets dimmer.

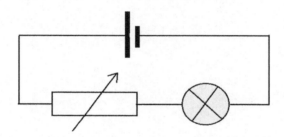

re will be lots of these in your house. The knobs on your electric cooker are variable resistors as are
volume and tone knobs on your electric guitar.

u may have noticed that there are three connections on this component. That is because it is actually
xed resistance with a sliding contact in the middle.

e just use contacts A and B then it is just a variable resistor as described above. When the slider B
ves to the right the resistance gets bigger.

3 contacts are used in a circuit called a potential divider which will be discussed in the next section.
tarists refer to these devices as "pots". The volume pot on my amp goes up to 11!

Summary of section 6

The resistance of a diode may be very big or very small depending on which way round it is connected

The resistance of a thermistor depends on its temperature

The resistance of an n.t.c. thermistor gets smaller when it gets hotter

The resistance of an LDR gets smaller when more light falls on it

The resistance of a variable resistor or rheostat depends on the position of a slider or knob.

Potential Dividers

sider these two resistors in series

What will the voltage across each resistor be?

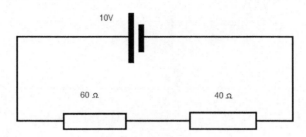

total resistance in the circuit is 100Ω so the current through both resistors will be

$$= \frac{V}{R} \qquad = 10/100 \qquad = 0.1A$$

voltage across the 60Ω resistor will therefore be V = I R = 0.1 x 60 = 6V

the voltage across the 40Ω resistor will be V = I R = 0.1 x 40 = 4V

 + 4 = 10V we can be confident that our answers are correct

re is a quicker way of figuring out what the voltages will be.

60Ω resistor is going to get 60/100 x the total voltage and the 40Ω resistor will get 40/100 x 10V

eneral if we have two resistors R_1 and R_2 in series then the voltage across R_1 is given by

$$V = V_{total} \times \left(\frac{R_1}{R_1 + R_2} \right)$$

e's another example, with some less friendly numbers, as this is quite an important concept

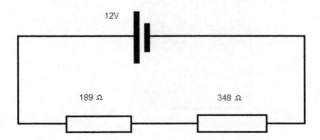

For the 189Ω resistor   V = 12 x (189 / 537) = 4.22V   The 348Ω resistor therefore gets 12 – 4.22 = 7.78V

This kid of circuit is a called a **potential divider**. The potential difference supplied by the power supply is divided between the two resistors.

Now imagine we replaced the first resistor with a thermistor and put a voltmeter across the second resistor

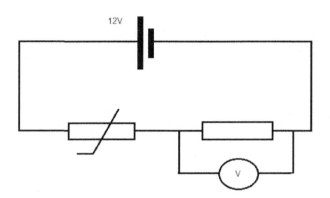

What would happen to the voltmeter reading if the thermistor was hotter?

Well if it were hotter the resistance of the thermistor would decrease. It would therefore get a smaller share of the supply voltage. The fixed resistor would get a bigger share and so the voltmeter reading wou increase.

We have made a device which gives a voltage reading which increases with temperature. Let's put this in little black box with a switch and an LED. We could then flog them on eBay or maybe even go on Dragons Den.

Who might want to buy it?

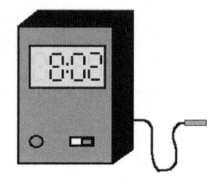

nyone who wants to measure temperature might want to buy our device, from scientists to cooks. We
ave built a temperature sensor.

ave a good look at the circuit below

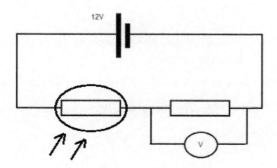

What would happen to the voltmeter reading if we shined a torch on the LDR?

What kind of sensor have we built now?

Who might find this device useful?

ow we have a light sensor. When light falls on the LDR its resistance gets less so the fixed resistor gets a
gger share of the voltage and the voltmeter reading increases. Gardeners, cricket umpires and
notographers will be lining up to buy this.

wo ways to produce a varying voltage

:udy these two circuits

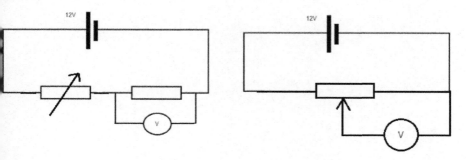

3

In both circuits we can produce a voltage smaller than the supply voltage. But, apart from being a bit more complicated to set up, the second circuit is better.

It can supply any voltage between 0 and 12V whereas the first circuit will not be able to give 0V

The output of the second circuit will also probably be more linear, i.e. if the slider is halfway we will get 6V

This is called a potentiometer circuit

Summary of section 7

A potential divider is a series circuit containing two or more resistors

The voltage of the power supply is shared between them

The voltage across each resistor is proportional to its fraction of the total resistance

If the resistance of one of the resistors depends on some external quantity, e.g. light intensity or temperature, the circuit can be used as a sensor

Such as circuit produces a voltage output which varies with the quantity which changes

One can step down the voltage from a power supply using a potential divider circuit

A potentiometer uses a resistor with a sliding contact to do this

8. Internal Resistance

agine you get a 6V bulb for your birthday. You discover pretty soon that it's not much use without a wer supply so you go to the local hardware store to buy a 6V battery. You bring a voltmeter with you to eck that you are getting the 6V you are paying for.

the battery isle you find a suitable battery labelled 6V and, when you test it with your voltmeter, you get reading of 6V. Hurrah! You buy it.

hen you get home you build the circuit below and tremble with excitement as you flick the switch closed.

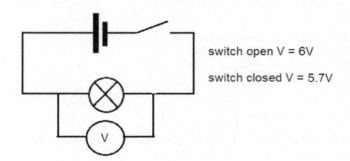

switch open V = 6V

switch closed V = 5.7V

your horror the voltmeter reading goes down to 5.7V when you close the switch. WHY?

ery component has some resistance (unless it's a superconductor). Power supplies have resistance too. ually it's quite small and in most questions you will ignore it. You often see the phrase "a power supply negligible internal resistance". In some questions, however, you will have to take it into account.

hink of the power supply's resistance as an extra resistance in series with the circuit. The 6V produced by e power supply is therefore shared between this internal resistance and the external circuit. In the ample above the external circuit, i.e. the bulb, gets 5.7V and the internal resistance gets 0.3V.

t's define some terms:

Quantity	Symbol	Units	Description
e e.m.f. of the power supply	ε	V	This is how much energy the power supply gives to every coulomb of charge that passes through it
e terminal p.d.	V	V	This is the voltage across the external circuit. It is the amount of energy transferred per coulomb by the bulb in the example above.
e lost volts	v	V	The energy transferred per coulomb **inside** the power supply due to its own resistance
rrent	I	A	The current in the circuit. Note this only flows when the switch is closed.
e resistance of the external cuit	R	Ω	The resistance of the bulb in this example
e internal resistance of the wer supply	r	Ω	The resistance of the power supply

How are these quantities related?

$\mathcal{E} = V + v$ from conservation of energy.

$V = IR$ you may have seen this before

$v = Ir$ similar to the above.

Note that if I = 0 then no volts are lost and the voltmeter in the example will tell us the e.m.f. when the switch is open

The equation you are given on the formula sheet is $\mathcal{E} = I(R + r).$

If you multiply this out you should see that we get the first equation on our list above.

Using the internal resistance equation

The voltage across a power supply falls from 6V to 5.7V when connected across a 2Ω resistor.

   a) *State the e.m.f. of the power supply*
   b) *Calculate the current that flows when the switch is closed*
   c) *Calculate the internal resistance of the power supply*

The e.m.f. is the voltage across the power supply when no current flows so $\mathcal{E}$ = 6V

$$I = \frac{V}{R} = \frac{5.7}{2} = 2.85A$$

$$r = \frac{(\mathcal{E}-V)}{I} \quad = 0.3 / 2.85 = 0.105Ω$$

More about internal resistance

A 1.5V AA battery has an internal resistance of about 1.5Ω

A 12V car battery has an internal resistance of 20mΩ

Your car's starter motor draws a current of about 100A from the car battery.

Could you start your car using 8 AA batteries in series?

No. Imagine we short circuited each battery, i.e. we connected a thick copper wire across its terminals. T only resistance in the circuit would be the internal resistance of the supply and we could work out the maximum current.

For the 8 x 1.5V AA batteries r = 8 x 1.5 = 12Ω   so I_{max} = 12/12 = 1A the car battery I_{max} = 12/0.02 = 600A

v to measure internal resistance of a power supply

:hod 1 (the easy way)

asure the e.m.f. of the supply with a voltmeter then connect a resistor across it and measure the ent through it and the p.d. across it.

:ulate the lost volts using $v = \mathcal{E} - V$ then $r = \dfrac{v}{I}$

:hod 2 (the clever more accurate way)

:d the circuit below

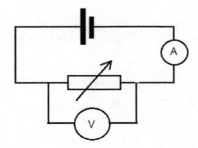

y the resistance of the rheostat and take pairs of readings of V and I then plot V against I and draw a of best fit as shown below.

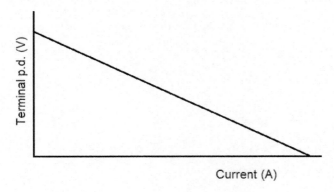

gradient of this graph = -r   Why?

$= V + Ir$ Re-arranging gives $V = -Ir + \mathcal{E}$

You should be familiar with the equation for a straight line $y = mx + c$ and comparing that with †
equation above tells us that $m = -r$

The intercept on the y axis, c, is equal to the e.m.f. and the x intercept is the maximum current.

Summary of section 8

Unless we are told that the power supply has "negligible internal resistance" then we need to take this in
account

Power supplies have a small resistance that we call their internal resistance

The e.m.f., $\mathcal{E}$, of a power supply is constant. It is the total energy transferred per coulomb that flows
through it. $\mathcal{E} = \dfrac{E}{Q}$

The e.m.f. of a power supply is the voltage across it when no current flows

The more current we draw from a power supply the less voltage we get from it

In situations where a large current is needed then the power supply should have a small internal
resistance

$$\mathcal{E} = I\,(R + r)$$

9. Resistivity

What does the resistance of a piece of wire depend on? (at constant T)

1. It will depend on its length

Doubling the length would be like putting 2 equal resistors in series so logically the resistance should be proportional to the length

2. It will depend on its cross sectional area

Doubling the cross sectional area would be like putting 2 equal resistors in parallel so logically the resistance should be inversely proportional to the cross sectional area

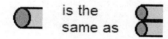 is the same as

3. It will depend on a property of the material that the wire is made of which we call its resistivity

$$\text{If } R \propto L \quad \text{and} \quad R \propto \frac{1}{A} \quad \text{then} \quad R \propto \frac{L}{A} \quad \text{or} \quad R = \text{constant} \times \frac{L}{A}$$

The resistivity of the material is defined as the constant in the equation for which we use the symbol ρ

Quantity	Symbol	Units	Description
Resistance	R	Ω	The resistance of the wire
Length	L	m	The length of the wire
Area	A	m^2	The cross sectional area of the wire
Resistivity	ρ	Ωm	The resistivity of the material the wire is made from

$$\text{So} \quad R = \rho \frac{L}{A} \quad \text{or} \quad \rho = \frac{R A}{L}$$

Using the resistivity equation

Aluminium has a resistivity of 2.65×10^{-8} Ωm

Calculate the resistance of an aluminium wire 2.5m long with a diameter of 0.6mm

$$A = \pi r^2 = \pi (0.3 \times 10^{-3})^2 = 2.83 \times 10^{-7} \, m^2$$

$$R = \rho \frac{L}{A} = 2.65 \times 10^{-8} \times \frac{2.5}{2.83 \times 10^{-7}} = 0.234 \Omega$$

Be careful working out the area. Express the radius in m not mm when you work it out so that your answer is in m^2 not mm^2

4

Finding the resistivity of a material

Remember that resistivity is a material property, i.e. it is not the property of an object but a property of the material that the object is made from. The resistance of an object will depend on its dimensions.

Another example is density and mass. One is a property of a material, the other is a property of an object.

Method 1 (The easy way)

- Get a wire, at least 1m long, of this material.
- Measure its length
- Measure its resistance using a multimeter.
- Measure its diameter using a micrometer and use this to calculate its cross sectional area.
- Calculate its resistivity.

Method 2 (The clever more accurate way)

- Attach (tape) the wire to a metre ruler making sure it is straight and without kinks.
- Attach a crocodile clip to the wire at the zero on the ruler. Another crocodile clip can be placed at different positions on the wire. Have it at 30.0cm initially.
- Measure the diameter of the wire three times along its length and calculate an average. Use this to calculate the cross sectional area.
- Build the circuit below and adjust the voltage of the power supply so that the ammeter reads 1A.

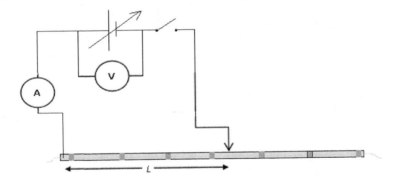

- Repeat the above for 40cm, 50cm etc up to 100cm changing the voltage each time so that the current is 1A.
- Calculate the resistance for each length of wire.
- Plot the wires resistance against its length and draw a line of best fit
- The gradient of this graph will equal $\dfrac{\rho}{A}$ so multiply by A to get ρ

Notes:

When we do a line of best fit we are not only averaging but we are also ignoring anomalies

If the current is the same for each length then the temperature of the wire should be the same

Conductivity

Just as conductance is the opposite of resistance there is a material property called conductivity which is, of course, the opposite of resistivity. Study the pairs of equations below and this should be clear.

$$R = \frac{V}{I} \quad G = \frac{I}{V} \qquad \rho = \frac{R A}{L} \quad \sigma = \frac{G L}{A} \qquad R = \rho \frac{L}{A} \quad G = \frac{\sigma A}{L}$$

Quantity	Symbol	Units	Description
Resistance	R	Ω	The resistance of the wire
Resistivity	ρ	Ωm	The resistivity of the material the wire is made from
Conductance	G	S	The conductance of the wire
Conductivity	σ	Sm^{-1}	The conductivity of the material the wire is made from

There are situations where it is more useful to think in terms of conductance and conductivity. One example is when you study magnetic circuits. One can draw a useful analogy between these and the ability of magnetic flux to flow around a magnetic circuit or through a material.

Summary of section 9

The resistance of a wire depends on the resistivity of the material it is made from

Plastic has a high resistivity   Copper has a low resistivity

$$\rho = \frac{R\,A}{L} \qquad R = \rho\,\frac{L}{A}$$

Conductivity is the opposite of resistivity

$$\sigma = \frac{G\,L}{A} \qquad G = \frac{\sigma\,A}{L}$$

estions

1. A student connects a bulb across a 6V power supply with negligible internal resistance.

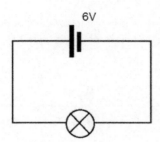

a) Add a voltmeter to the circuit to measure the voltage across the bulb
b) Add an ammeter to the circuit to measure the current through the bulb

current is measured to be 0.12A

c) Calculate the resistance of the bulb

d) How much charge flows through the bulb every minute?

e) How many electrons carry this charge?

f) Calculate the power of the bulb

g) Calculate the energy transferred by the bulb every minute

h) On the axes below sketch a typical V I characteristic for a bulb

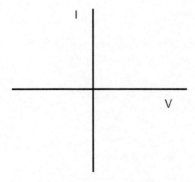

i) Describe how the resistance of the bulb changes when we increase the voltage across it

j) Explain why this happens

2. A bulb and a resistor are connected in series across a 6V power supply of negligible internal resistance. The voltage across the bulb is measured to be 3.3V and the current through it is 42m

6V

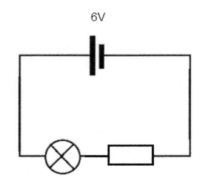

a) Calculate the resistance of the bulb

b) Calculate the voltage across the fixed resistor

c) Calculate its resistance

An identical bulb is now added to the circuit in parallel with the first.

d) Draw this on the circuit above

e) Will the following quantities increase, decrease or stay the same?
i)   The resistance of the circuit

ii)   The current drawn from the supply

iii)   The voltage across the fixed resistor

3. Calculate the resistance of the combinations of resistors below

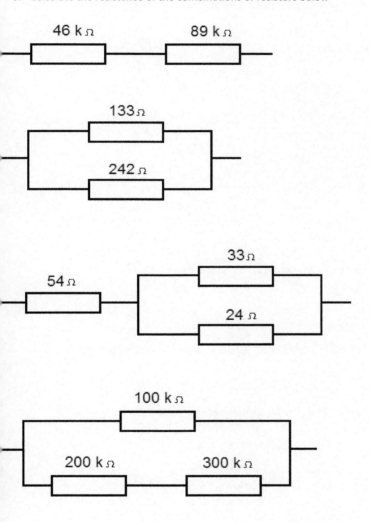

46 kΩ   89 kΩ

133 Ω

242 Ω

54 Ω

33 Ω

24 Ω

100 kΩ

200 kΩ   300 kΩ

student needs a resistance of 450Ω but only has 100Ω resistors available

ow might these be arranged to produce the desired resistance?

4. The resistance of a particular thermistor varies with temperature as shown below. It is connected in series with a 100Ω fixed resistor and a 10V power supply with negligible internal resistance

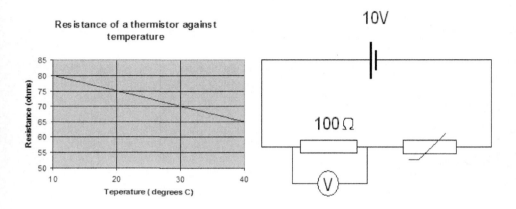

Resistance of a thermistor against temperature

10V

100 Ω

a) State the resistance of the thermistor at 10, 20, 30 and 40 degrees C

b) Calculate the voltage across the fixed resistor at each of these temperatures

c) Sketch the relationship between this voltage and the temperature

d) Describe a use for this circuit

A battery of e.m.f. 12V is connected across a 2Ω motor. The voltage across the motor is measured as 11.3V. Calculate the internal resistance of the cell

A cell of e.m.f. 6V and internal resistance 0.1Ω is connected across a bulb with a working resistance of 3Ω. What will be the voltage across the bulb?

What will be the current through it?

Six 1.5V cells, each with an internal resistance of 0.2Ω, are connected in series. What is the e.m.f. and internal resistance of this combination?

If it were connected across a 10Ω heater what current and p.d. would they deliver?

8. A copper wire has a length of 20m and a diameter of 0.5mm. The resistivity of copper is $1.7 \times 10^{-8}\,\Omega m$.

a) Calculate the conductivity of copper

b) Calculate the cross sectional area of the wire

c) Calculate the resistance of the copper wire.

9. A 1.9 cm length of tungsten filament in a small light bulb has a working resistance of $670\,\Omega$. The resistivity of tungsten is $5.6 \times 10^{-8}\,\Omega m$.

a) What current will flow through the bulb when 240V is put across it?

b) Calculate the power of the bulb

c) Calculate the cross sectional area of the filament.

d) If the filament has a circular cross section calculate its diameter

10. The graphite in a pencil is 15cm long and has a diameter of 1.1mm. When 2V is put across the pencil a current of 1.29A flows through it.

a) Calculate the resistance of the graphite core

b) Calculate the cross sectional area of the graphite core

c) Calculate the resistivity of graphite

swers to numerical questions

1.

c) R = V / I = 6/1.2 = 50Ω

d) Q = I t =0.12 x 60 = 7.2C
e) 7.2 / 1.6 x 10^{-19} = 4.5 x 10^{19}
f) P = V I = 6 x 0.12 = 0.72W
g) E = Pt = 0.72 x 60 = 43.2J

R = /I = 3.3/42x10^{-3} = 78.6Ω   b) 6-3.3 = 2.7V

R = V/I = 2.7/42x10^{-3} = 64.3Ω   e) i) decrease ii) increase iii) increase

5kΩ   85.8Ω   67.9Ω   83.3kΩ   4 in series + 2 in parallel

10	20	30	40
80	75	70	65
5.5	5.71	5.88	6.0

V/R = 11.3/2 = 5.65A   r = (Ɛ-V)/I = 0.7/5.65 = 0.124Ω

Ɛ/(R + r) = 6/3.1 = 1.94A   V = IR = 1.94 x 3 = 5.82V

6 x 1.5 = 9V   r=6 x 0.1 = 1.2Ω   I = 9/(1.2 + 10) = 0.804A   V = IR = 8.04V

= 1/ρ = 1/1.7x10^{-8} = 58.8 x 10^6 Sm^{-1} A = 1.96 x $10^{-7}m^2$ R = 1.73Ω

0.358A   P = 86W   A = 1.59 x $10^{-12}m^2$ d = 7.1 x 10^{-7}m

.

= 1.55Ω   A = 9.5 x $10^{-7}m^2$ P = 9.82 x 10^{-6} Ωm

Printed in Great Britain
by Amazon